This
Password book
belongs to

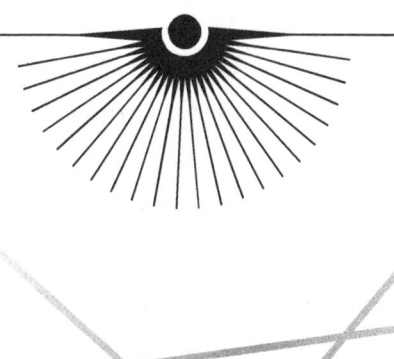

© Copyright 2021 - All rights reserved.

-You may not reproduce, duplicate, or send the contents of this book without direct written permission from the author. You cannot hereby despite any circumstance blame the publisher or hold him or her to legal responsibility for any reparation, compensations, or monetary forfeiture owing to the information included herein, either in a direct or an indirect way.

-Legal Notice: This book has copyright protection. You can use the book for personal purposes. You should not sell, use, alter, distribute, quote, take excerpts, or paraphrase in part or whole the material contained in this book without obtaining the permission of the author first.

-Disclaimer Notice: You must take note that the information in this document is for casual reading and entertainment purposes only. We have made every attempt to provide accurate, up-to-date, and reliable information. We do not express or imply guarantees of any kind. The persons who read admit that the writer is not occupied in giving legal, financial, medical, or other advice. We put this book content by sourcing various places.

-Please consult a licensed professional before you try any techniques shown in this book. By going through this document, the book lover comes to an agreement that under no situation is the author accountable for any forfeiture, direct or indirect, which they may incur because of the use of material contained in this document, including, but not limited to, — errors, omissions, or inaccuracies

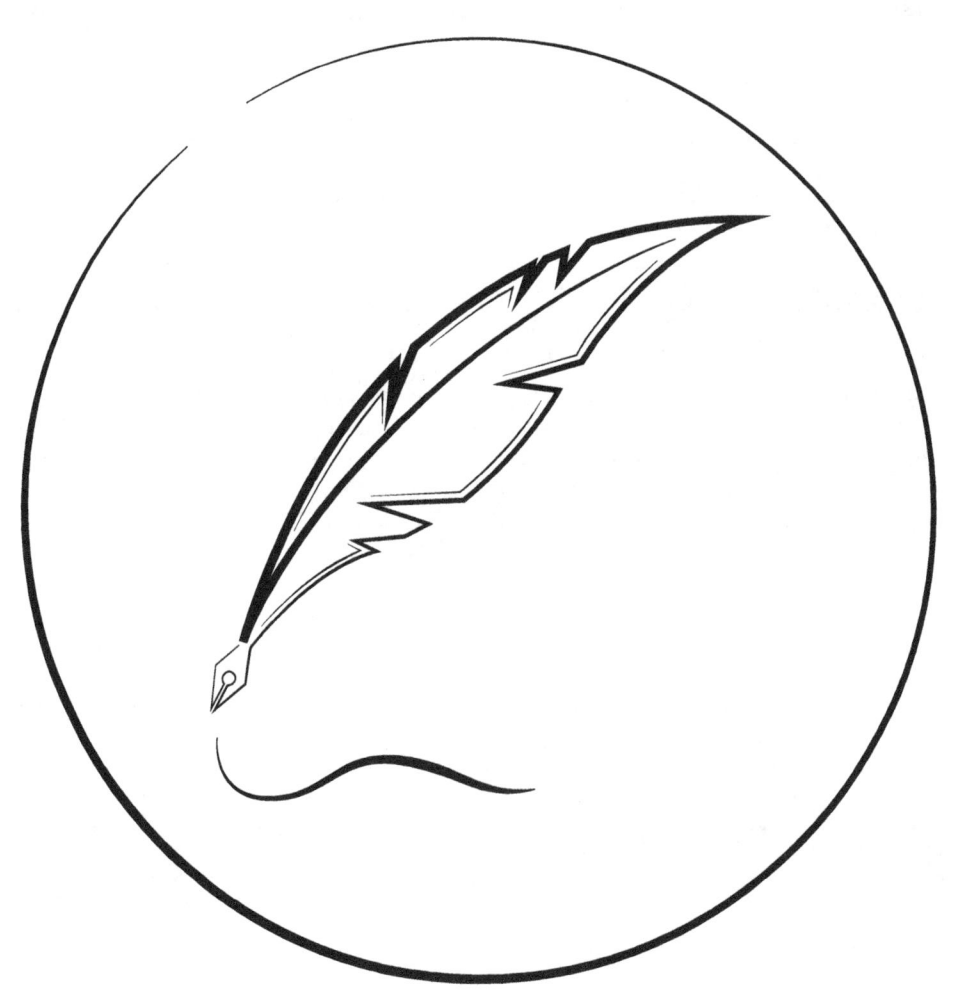

A

Web site: ....

Email/phone no: ....
Username: ....
Password: ....
Notes: ....

....
....
....

Web site: ....

Email/phone no: ....
Username: ....
Password: ....
Notes: ....

....
....
....

Web site: ....

Email/phone no: ....
Username: ....
Password: ....
Notes: ....

....
....
....

Web site: .... A

Email/phone no: ....
Username: ....
Password: ....
Notes: ....

....
....
....

Web site: ....

Email/phone no: ....
Username: ....
Password: ....
Notes: ....

....
....
....

Web site: ....

Email/phone no: ....
Username: ....
Password: ....
Notes: ....

....
....
....

A

Web site: ....

Email/phone no: ....

Username: ....

Password: ....

Notes: ....

....

....

....

Web site: ....

Email/phone no: ....

Username: ....

Password: ....

Notes: ....

....

....

....

Web site: ....

Email/phone no: ....

Username: ....

Password: ....

Notes: ....

....

....

....

Web site:... A

Email/phone no:..
Username:..
Password:..
Notes:...

....
....
....

Web site:...

Email/phone no:..
Username:..
Password:..
Notes:...

....
....
....

Web site:...

Email/phone no:..
Username:..
Password:..
Notes:...

....
....
....

Web site:...

Email / phone no:..
Username:..

Password:..

Notes:...

....
....
....

Web site:...

Email / phone no:..
Username:..

Password:..

Notes:...

....
....
....

Web site:...

Email / phone no:..
Username:..

Password:..

Notes:...

....
....
....

Web site:..

Email/phone no:..
Username:..
Password:...
Notes:...

....
....
....

Web site:..

Email/phone no:..
Username:..
Password:...
Notes:...

....
....
....

Web site:..

Email/phone no:..
Username:..
Password:...
Notes:...

....
....
....

Web site: ....

Email / phone no: ....

Username: ....

Password: ....

Notes: ....

....

....

....

Web site: ....

Email / phone no: ....

Username: ....

Password: ....

Notes: ....

....

....

....

Web site: ....

Email / phone no: ....

Username: ....

Password: ....

Notes: ....

....

....

....

Web site:

Email/phone no:
Username:
Password:
Notes:

Web site:

Email/phone no:
Username:
Password:
Notes:

Web site:

Email/phone no:
Username:
Password:
Notes:

Web site: ....

Email/phone no: ....
Username: ....

Password: ....

Notes: ....

....

....

....

Web site: ....

Email/phone no: ....

Username: ....

Password: ....

Notes: ....

....

....

....

Web site: ....

Email/phone no: ....

Username: ....

Password: ....

Notes: ....

....

....

....

Web site:..

Email/phone no:...

Username:..

Password:..

Notes:...

....

....

....

Web site:..

Email/phone no:...

Username:..

Password:..

Notes:...

....

....

....

Web site:..

Email/phone no:...

Username:..

Password:..

Notes:...

....

....

....

Web site:..

Email/phone no:..
Username:...

Password:...

Notes:..

....

....

....

Web site:..

Email/phone no:..

Username:...

Password:...

Notes:..

....

....

....

Web site:..

Email/phone no:..

Username:...

Password:...

Notes:..

....

....

....

Web site: ....

Email / phone no: ....
Username: ....

Password: ....

Notes: ....

....
....
....

Web site: ....

Email / phone no: ....
Username: ....

Password: ....

Notes: ....

....
....
....

Web site: ....

Email / phone no: ....
Username: ....

Password: ....

Notes: ....

....
....
....

Web site:..

Email/phone no:...
Username:...
Password:...
Notes:..

....
....
....

Web site:..

Email/phone no:...
Username:...
Password:...
Notes:..

....
....
....

Web site:..

Email/phone no:...
Username:...
Password:...
Notes:..

....
....
....

Web site:..

Email/phone no:..
Username:..
Password:...
Notes:..

....
....
....

Web site:..

Email/phone no:..
Username:..
Password:...
Notes:..

....
....
....

Web site:..

Email/phone no:..
Username:..
Password:...
Notes:..

....
....
....

Web site:..

Email/phone no:..
Username:..
Password:...
Notes:..

....
....
....

Web site:..

Email/phone no:..
Username:..
Password:...
Notes:..

....
....
....

Web site:..

Email/phone no:..
Username:..
Password:...
Notes:..

....
....
....

Web site:..

Email/phone no:..
Username:...
Password:..
Notes:..

....
....
....

Web site:..

Email/phone no:..
Username:...
Password:..
Notes:..

....
....
....

Web site:..

Email/phone no:..
Username:...
Password:..
Notes:..

....
....
....

Web site:..

Email/phone no:..
Username:...
Password:..
Notes:..

....
....
....

Web site:..

Email/phone no:..
Username:...
Password:..
Notes:..

....
....
....

Web site:..

Email/phone no:..
Username:...
Password:..
Notes:..

....
....
....

Web site:

Email/phone no:
Username:
Password:
Notes:

....
....
....

Web site:

Email/phone no:
Username:
Password:
Notes:

....
....
....

Web site:

Email/phone no:
Username:
Password:
Notes:

....
....
....

Web site:..

Email/phone no:..
Username:...
Password:...
Notes:..

....
....
....

Web site:..

Email/phone no:..
Username:...
Password:...
Notes:..

....
....
....

Web site:..

Email/phone no:..
Username:...
Password:...
Notes:..

....
....
....

Web site:...

Email / phone no:..
Username:...

Password:...

Notes:..

....
....
....

Web site:...

Email / phone no:..
Username:...

Password:...

Notes:..

....
....
....

Web site:...

Email / phone no:..

Username:...

Password:...

Notes:..

....
....
....

Web site: ....

Email/phone no: ....
Username: ....
Password: ....
Notes: ....
....
....
....

Web site: ....

Email/phone no: ....
Username: ....
Password: ....
Notes: ....
....
....
....

Web site: ....

Email/phone no: ....
Username: ....
Password: ....
Notes: ....
....
....
....

Web site:..

Email/phone no:..
Username:..
Password:..
Notes:..
....
....
....

Web site:..

Email/phone no:..
Username:..
Password:..
Notes:..
....
....
....

Web site:..

Email/phone no:..
Username:..
Password:..
Notes:..
....
....
....

Web site:..

Email/phone no:...
Username:..
Password:...
Notes:..
....
....
....

Web site:..

Email/phone no:...
Username:..
Password:...
Notes:..
....
....
....

Web site:..

Email/phone no:...
Username:..
Password:...
Notes:..
....
....
....

Web site:..

Email/phone no: ....
Username:. ....
Password:. ....
Notes:. ....

....
....
....

Web site:..

Email/phone no: ....
Username:. ....
Password:. ....
Notes:. ....

....
....
....

Web site:..

Email/phone no: ....
Username:. ....
Password:. ....
Notes:. ....

....
....
....

F

Web site: ....

Email/phone no: ....

Username: ....

Password: ....

Notes: ....

....

....

....

Web site: ....

Email/phone no: ....

Username: ....

Password: ....

Notes: ....

....

....

....

Web site: ....

Email/phone no: ....

Username: ....

Password: ....

Notes: ....

....

....

....

Web site:

Email/phone no:
Username:
Password:
Notes:

Web site:

Email/phone no:
Username:
Password:
Notes:

Web site:

Email/phone no:
Username:
Password:
Notes:

Web site:...

Email/phone no:...

Username:..

Password:...

Notes:...

....

....

....

Web site:...

Email/phone no:...

Username:..

Password:...

Notes:...

....

....

....

Web site:...

Email/phone no:...

Username:..

Password:...

Notes:...

....

....

....

Web site:

Email/phone no:
Username:
Password:
Notes:

Web site:

Email/phone no:
Username:
Password:
Notes:

Web site:

Email/phone no:
Username:
Password:
Notes:

Web site:

Email/phone no:
Username:
Password:
Notes:

Web site:

Email/phone no:
Username:
Password:
Notes:

Web site:

Email/phone no:
Username:
Password:
Notes:

Web site:..

Email/phone no:...
Username:..
Password:..
Notes:..

....
....
....

Web site:..

Email/phone no:...
Username:..
Password:..
Notes:..

....
....
....

Web site:..

Email/phone no:...
Username:..
Password:..
Notes:..

....
....
....

Web site:...

Email/phone no:..
Username:...
Password:..
Notes:..

....
....
....

Web site:...

Email/phone no:..
Username:...
Password:..
Notes:..

....
....
....

Web site:...

Email/phone no:..
Username:...
Password:..
Notes:..

....
....
....

Web site: ....

Email/phone no: ....
Username: ....
Password: ....
Notes: ....

....
....
....

Web site: ....

Email/phone no: ....
Username: ....
Password: ....
Notes: ....

....
....
....

Web site: ....

Email/phone no: ....
Username: ....
Password: ....
Notes: ....

....
....
....

Web site:..

Email/phone no:..
Username:..
Password:..
Notes:..
....
....
....

Web site:..

Email/phone no:..
Username:..
Password:..
Notes:..
....
....
....

Web site:..

Email/phone no:..
Username:..
Password:..
Notes:..
....
....
....

Web site:..

Email/phone no:..
Username:..
Password:..
Notes:..

....
....
....

Web site:..

Email/phone no:..
Username:..
Password:..
Notes:..

....
....
....

Web site:..

Email/phone no:..
Username:..
Password:..
Notes:..

....
....
....

Web site:..

Email/phone no:...
Username:..
Password:..
Notes:..

....
....
....

Web site:..

Email/phone no:...
Username:..
Password:..
Notes:..

....
....
....

Web site:..

Email/phone no:...
Username:..
Password:..
Notes:..

....
....
....

Web site:..

Email/phone no:..
Username:..
Password:..
Notes:...

....
....
....

Web site:..

Email/phone no:..
Username:..
Password:..
Notes:...

....
....
....

Web site:..

Email/phone no:..
Username:..
Password:..
Notes:...

....
....
....

Web site: ....

Email/phone no: ....
Username: ....
Password: ....
Notes: ....

....
....
....

Web site: ....

Email/phone no: ....
Username: ....
Password: ....
Notes: ....

....
....
....

Web site: ....

Email/phone no: ....
Username: ....
Password: ....
Notes: ....

....
....
....

Web site:..

Email/phone no:..
Username:..
Password:..
Notes:...

....
....
....

Web site:.. J

Email/phone no:..
Username:..
Password:..
Notes:...

....
....
....

Web site:..

Email/phone no:..
Username:..
Password:..
Notes:...

....
....
....

Web site:

Email/phone no:
Username:
Password:
Notes:

Web site:

Email/phone no:
Username:
Password:
Notes:

Web site:

Email/phone no:
Username:
Password:
Notes:

Web site: ....

Email/phone no: ....
Username: ....
Password: ....
Notes: ....

....
....
....

Web site: ....

Email/phone no: ....
Username: ....
Password: ....
Notes: ....

....
....
....

Web site: ....

Email/phone no: ....
Username: ....
Password: ....
Notes: ....

....
....
....

Web site:..

Email/phone no:..
Username:...
Password:..
Notes:..

....
....
....

Web site:..

Email/phone no:..
Username:...
Password:..
Notes:..

....
....
....

Web site:..

Email/phone no:..
Username:...
Password:..
Notes:..

....
....
....

Web site:

Email/phone no:
Username:
Password:
Notes:

....
....
....

Web site:

Email/phone no:
Username:
Password:
Notes:

....
....
....

Web site:

Email/phone no:
Username:
Password:
Notes:

....
....
....

Web site:

Email/phone no:
Username:
Password:
Notes:

Web site:

Email/phone no:
Username:
Password:
Notes:

Web site:

Email/phone no:
Username:
Password:
Notes:

Web site:

Email/phone no:
Username:
Password:
Notes:
....
....
....

Web site:

Email/phone no:
Username:
Password:
Notes:
....
....
....

Web site:

Email/phone no:
Username:
Password:
Notes:
....
....
....

Web site:

Email/phone no:

Username:

Password:

Notes:

....

....

....

Web site:

Email/phone no:

Username:

Password:

Notes:

....

....

....

Web site:

Email/phone no:

Username:

Password:

Notes:

....

....

....

Web site:

Email/phone no:
Username:
Password:
Notes:

....
....
....

Web site:

Email/phone no:
Username:
Password:
Notes:

....
....
....

Web site:

Email/phone no:
Username:
Password:
Notes:

....
....
....

Web site:

Email/phone no:
Username:
Password:
Notes:

...
...
...

Web site:

Email/phone no:
Username:
Password:
Notes:

...
...
...

Web site:

Email/phone no:
Username:
Password:
Notes:

...
...
...

Web site:..

Email/phone no:..
Username:..
Password:..
Notes:..

....
....
....

Web site:..

Email/phone no:..
Username:..
Password:..
Notes:..

....
....
....

Web site:..

Email/phone no:..
Username:..
Password:..
Notes:..

....
....
....

Web site:...

Email / phone no:..
Username:...
Password:..
Notes:...

....
....
....

Web site:...

Email / phone no:..
Username:...
Password:..
Notes:...

....
....
....

Web site:...

Email / phone no:..
Username:...
Password:..
Notes:...

....
....
....

Web site:..

Email/phone no:..
Username:...
Password:..
Notes:..

....
....
....

Web site:..

Email/phone no:..
Username:...
Password:..
Notes:..

....
....
....

Web site:..

Email/phone no:..
Username:...
Password:..
Notes:..

....
....
....

Web site: ....

Email/phone no: ....

Username: ....

Password: ....

Notes: ....

....

....

....

Web site: ....

Email/phone no: ....

Username: ....

Password: ....

Notes: ....

....

....

....

Web site: ....

Email/phone no: ....

Username: ....

Password: ....

Notes: ....

....

....

....

Web site:..

Email/phone no:..
Username:..
Password:..
Notes:..

....
....
....

Web site:..

Email/phone no:..
Username:..
Password:..
Notes:..

....
....
....

Web site:..

Email/phone no:..
Username:..
Password:..
Notes:..

....
....
....

Web site:..

Email / phone no:..
Username:..
Password:..
Notes:...
....
....
....

Web site:..

Email / phone no:..
Username:..
Password:..
Notes:...
....
....
....

Web site:..

Email / phone no:..
Username:..
Password:..
Notes:...
....
....
....

Web site:

Email/phone no:
Username:
Password:
Notes:

....
....
....

Web site:

Email/phone no:
Username:
Password:
Notes:

....
....
....

Web site:

Email/phone no:
Username:
Password:
Notes:

....
....
....

Web site: ....

Email / phone no: ....
Username: ....
Password: ....
Notes: ....

....
....
....

Web site: ....

Email / phone no: ....
Username: ....
Password: ....
Notes: ....

....
....
....

Web site: ....

Email / phone no: ....
Username: ....
Password: ....
Notes: ....

....
....
....

Web site:

Email/phone no:
Username:
Password:
Notes:

....
....
....

Web site:

Email/phone no:
Username:
Password:
Notes:

....
....
....

Web site:

Email/phone no:
Username:
Password:
Notes:

....
....
....

Web site:..

Email/phone no:..
Username:...
Password:...
Notes:..
....
....
....

Web site:..

Email/phone no:..
Username:...
Password:...
Notes:..
....
....
....

Web site:..

Email/phone no:..
Username:...
Password:...
Notes:..
....
....
....

Web site:..

Email/phone no:..
Username:..
Password:..
Notes:..

....
....
....

Web site:..

Email/phone no:..
Username:..
Password:..
Notes:..

....
....
....

Web site:..

Email/phone no:..
Username:..
Password:..
Notes:..

....
....
....

Web site:...

Email/phone no:...
Username:..
Password:..
Notes:..

....
....
....

Web site:...

Email/phone no:...
Username:..
Password:..
Notes:..

....
....
....

Web site:...

Email/phone no:...
Username:..
Password:..
Notes:..

....
....
....

Web site:

Email/phone no:
Username:
Password:
Notes:

....
....
....

Web site:

Email/phone no:
Username:
Password:
Notes:

....
....
....

Web site:

Email/phone no:
Username:
Password:
Notes:

....
....
....

Web site:

Email/phone no:

Username:

Password:

Notes:

....

....

....

Web site:

Email/phone no:

Username:

Password:

Notes:

....

....

....

Web site:

Email/phone no:

Username:

Password:

Notes:

....

....

....

Web site:

Email/phone no:
Username:
Password:
Notes:

Web site:

Email/phone no:
Username:
Password:
Notes:

Web site:

Email/phone no:
Username:
Password:
Notes:

Web site:..

Email/phone no:..
Username:...
Password:...
Notes:...

....
....
....

Web site:..

Email/phone no:..
Username:...
Password:...
Notes:...

....
....
....

Web site:..

Email/phone no:..
Username:...
Password:...
Notes:...

....
....
....

Web site:...

Email/phone no:...
Username:..
Password:...
Notes:..

....
....
....

Web site:...

Email/phone no:...
Username:..
Password:...
Notes:..

....
....
....

Web site:...

Email/phone no:...
Username:..
Password:...
Notes:..

....
....
....

Web site:

Email / phone no:
Username:
Password:
Notes:

Web site:

Email / phone no:
Username:
Password:
Notes:

Web site:

Email / phone no:
Username:
Password:
Notes:

Web site:..

Email / phone no:..
Username:...
Password:...
Notes:..

....
....
....

Web site:..

Email / phone no:..
Username:...
Password:...
Notes:..

....
....
....

Web site:..

Email / phone no:..
Username:...
Password:...
Notes:..

....
....
....

Web site:

Email/phone no:
Username:
Password:
Notes:

Web site:

Email/phone no:
Username:
Password:
Notes:

Web site:

Email/phone no:
Username:
Password:
Notes:

Web site:

Email/phone no:

Username:

Password:

Notes:

....

....

....

Web site:

Email/phone no:

Username:

Password:

Notes:

....

....

....

Web site:

Email/phone no:

Username:

Password:

Notes:

....

....

....

Web site:

Email / phone no: ....
Username: ....
Password: ....
Notes: ....

....
....
....

Web site:

Email / phone no: ....
Username: ....
Password: ....
Notes: ....

....
....
....

Web site:

Email / phone no: ....
Username: ....
Password: ....
Notes: ....

....
....
....

Web site:..

Email/phone no:...
Username:...
Password:...
Notes:..

....
....
....

Web site:..

Email/phone no:...
Username:...
Password:...
Notes:..

....
....
....

Web site:..

Email/phone no:...
Username:...
Password:...
Notes:..

....
....
....

Web site:..

Email/phone no:..
Username:..
Password:...
Notes:...

....
....
....

Web site:..

Email/phone no:..
Username:..
Password:...
Notes:...

....
....
....

Web site:..

Email/phone no:..
Username:..
Password:...
Notes:...

....
....
....

Web site:..

Email/phone no:..
Username:..
Password:..
Notes:...
....
....
....
....

Web site:..

Email/phone no:..
Username:..
Password:..
Notes:...
....
....
....
....

Web site:..

Email/phone no:..
Username:..
Password:..
Notes:...
....
....
....

Web site:..

Email/phone no:..
Username:..
Password:...
Notes:..

....
....
....

Web site:..

Email/phone no:..
Username:..
Password:...
Notes:..

....
....
....

Web site:..

Email/phone no:..
Username:..
Password:...
Notes:..

....
....
....

Web site:

Email/phone no:
Username:
Password:
Notes:

Web site:

Email/phone no:
Username:
Password:
Notes:

Web site:

Email/phone no:
Username:
Password:
Notes:

Web site:..

Email/phone no:..
Username:..
Password:...
Notes:..

....
....
....

Web site:..

Email/phone no:..
Username:..
Password:...
Notes:..

....
....
....

Web site:..

Email/phone no:..
Username:..
Password:...
Notes:..

....
....
....

Web site:...

Email/phone no:...
Username:..
Password:..
Notes:..

....
....
....

Web site:...

Email/phone no:...
Username:..
Password:..
Notes:..

....
....
....

Web site:...

Email/phone no:...
Username:..
Password:..
Notes:..

....
....
....

Web site:..

Email/phone no:..
Username:..
Password:..
Notes:..

....
....
....

Web site:..

Email/phone no:..
Username:..
Password:..
Notes:..

....
....
....

Web site:..

Email/phone no:..
Username:..
Password:..
Notes:..

....
....
....

Web site: ....

Email / phone no: ....
Username: ....
Password: ....
Notes: ....

....
....
....

Web site: ....

Email / phone no: ....
Username: ....
Password: ....
Notes: ....

....
....
....

Web site: ....

Email / phone no: ....
Username: ....
Password: ....
Notes: ....

....
....
....

Web site: ....

Email/phone no: ....
Username: ....
Password: ....
Notes: ....

....
....
....

Web site: ....

Email/phone no: ....
Username: ....
Password: ....
Notes: ....

....
....
....

Web site: ....

Email/phone no: ....
Username: ....
Password: ....
Notes: ....

....
....
....

Web site:..

Email/phone no:...
Username:..
Password:...
Notes:..

....
....
....

Web site:..

Email/phone no:...
Username:..
Password:...
Notes:..

....
....
....

Web site:..

Email/phone no:...
Username:..
Password:...
Notes:..

....
....
....

T

Web site:..

Email/phone no:...
Username:..
Password:...
Notes:..
....
....
....

Web site:..

Email/phone no:...
Username:..
Password:...
Notes:..
....
....
....

Web site:..

Email/phone no:...
Username:..
Password:...
Notes:..
....
....
....

Web site:..

Email / phone no:...
Username:...
Password:..
Notes:...

....
....
....

Web site:..

Email / phone no:...
Username:...
Password:..
Notes:...

....
....
....

Web site:..

Email / phone no:...
Username:...
Password:..
Notes:...

....
....
....

Web site:..

Email/phone no:...
Username:..
Password:..
Notes:...

....
....
....

Web site:..

Email/phone no:...
Username:..
Password:..
Notes:...

....
....
....

Web site:..

Email/phone no:...
Username:..
Password:..
Notes:...

....
....
....

Web site: ....

Email / phone no: ....
Username: ....
Password: ....
Notes: ....

....
....
....

Web site: ....

Email / phone no: ....
Username: ....
Password: ....
Notes: ....

....
....
....

Web site: ....

Email / phone no: ....
Username: ....
Password: ....
Notes: ....

....
....
....

Web site:

Email/phone no:

Username:

Password:

Notes:

Web site:

Email/phone no:

Username:

Password:

Notes:

Web site:

Email/phone no:

Username:

Password:

Notes:

Web site:..

Email / phone no:..

Username:..

Password:...

Notes:...

....

....

....

Web site:..

Email / phone no:..

Username:..

Password:...

Notes:...

....

....

....

Web site:..

Email / phone no:..

Username:..

Password:...

Notes:...

....

....

....

Web site:..

Email/phone no:...
Username:..
Password:..
Notes:..

....
....
....

Web site:..

Email/phone no:...
Username:..
Password:..
Notes:..

....
....
....

Web site:..

Email/phone no:...
Username:..
Password:..
Notes:..

....
....
....

Web site:

Email/phone no:
Username:
Password:
Notes:
....
....
....

Web site:

Email/phone no:
Username:
Password:
Notes:
....
....
....

Web site:

Email/phone no:
Username:
Password:
Notes:
....
....
....

V

Web site:..

Email/phone no:..
Username:..
Password:..
Notes:..
....
....
....

Web site:..

Email/phone no:..
Username:..
Password:..
Notes:..
....
....
....

Web site:..

Email/phone no:..
Username:..
Password:..
Notes:..
....
....
....

Web site:..

Email / phone no:..
Username:..
Password:..
Notes:..
....
....
....

Web site:..

Email / phone no:..
Username:..
Password:..
Notes:..
....
....
....

Web site:..

Email / phone no:..
Username:..
Password:..
Notes:..
....
....
....

Web site:..

Email/phone no:..
Username:...
Password:...
Notes:..

....
....
....

Web site:..

Email/phone no:..
Username:...
Password:...
Notes:..

....
....
....

Web site:..

Email/phone no:..
Username:...
Password:...
Notes:..

....
....
....

Web site:..

Email/phone no:...
Username:..
Password:...
Notes:..

....
....
....

Web site:..

Email/phone no:...
Username:..
Password:...
Notes:..

....
....
....

Web site:..

Email/phone no:...
Username:..
Password:...
Notes:..

....
....
....

W

Web site:

Email / phone no:

Username:

Password:

Notes:

Web site:

Email / phone no:

Username:

Password:

Notes:

Web site:

Email / phone no:

Username:

Password:

Notes:

Web site:

Email/phone no:
Username:
Password:
Notes:
....
....
....

Web site:

Email/phone no:
Username:
Password:
Notes:
....
....
....

Web site:

Email/phone no:
Username:
Password:
Notes:
....
....
....

X

Web site:..

Email/phone no:...
Username:..
Password:...
Notes:..
....
....
....

Web site:..

Email/phone no:...
Username:..
Password:...
Notes:..
....
....
....

Web site:..

Email/phone no:...
Username:..
Password:...
Notes:..
....
....
....

Web site:..

Email/phone no:..
Username:...
Password:..
Notes:..

....
....
....

Web site:..

Email/phone no:..
Username:...
Password:..
Notes:..

....
....
....

Web site:..

Email/phone no:..
Username:...
Password:..
Notes:..

....
....
....

Web site:..

Email / phone no:..
Username:..
Password:...
Notes:...

....
....
....

Web site:..

Email / phone no:..
Username:..
Password:...
Notes:...

....
....
....

Web site:..

Email / phone no:..
Username:..
Password:...
Notes:...

....
....
....

Web site:..

Email/phone no:...
Username:...
Password:..
Notes:..

....
....
....

Web site:..

Email/phone no:...
Username:...
Password:..
Notes:..

....
....
....

Web site:..

Email/phone no:...
Username:...
Password:..
Notes:..

....
....
....

Web site:...

Email/phone no:...
Username:..
Password:...
Notes:..
....
....
....

Web site:...

Email/phone no:...
Username:..
Password:...
Notes:..
....
....
....

Web site:...

Email/phone no:...
Username:..
Password:...
Notes:..
....
....
....

Web site:...

Email/phone no:...
Username:..
Password:..
Notes:...
....
....
....

Web site:...

Email/phone no:...
Username:..
Password:..
Notes:...
....
....
....

Web site:...

Email/phone no:...
Username:..
Password:..
Notes:...
....
....
....

Web site:..

Email/phone no:..
Username:..
Password:..
Notes:..

....
....
....

Web site:..

Email/phone no:..
Username:..
Password:..
Notes:..

....
....
....

Web site:..

Email/phone no:..
Username:..
Password:..
Notes:..

....
....
....

Web site:..

Email/phone no:..
Username:..
Password:..
Notes:..

....
....
....

Web site:..

Email/phone no:..
Username:..
Password:..
Notes:..

....
....
....

Web site:..

Email/phone no:..
Username:..
Password:..
Notes:..

....
....
.... Z

Web site: ....

Email/phone no: ....
Username: ....
Password: ....
Notes: ....

....
....
....

Web site: ....

Email/phone no: ....
Username: ....
Password: ....
Notes: ....

....
....
....

Web site: ....

Email/phone no: ....
Username: ....
Password: ....
Notes: ....

....
....
....

Web site:..

Email/phone no:...
Username:..
Password:..
Notes:..
....
....
....

Web site:..

Email/phone no:...
Username:..
Password:..
Notes:..
....
....
....

Web site:..

Email/phone no:...
Username:..
Password:..
Notes:..
....
....
....

Notes

Notes

Notes

Notes

Thank you

We hope you enjoyed this journal. As a small family company, your feedback is very important to us. Please let us know how you like our book at:

alexruell.print@gmail.com